Beatriz Magalhães
Vladimiro Miranda

Reactive Power Planning

AF138577

Beatriz Magalhães
Vladimiro Miranda

Reactive Power Planning

LAP LAMBERT Academic Publishing

Impressum / Imprint

Bibliografische Information der Deutschen Nationalbibliothek: Die Deutsche Nationalbibliothek verzeichnet diese Publikation in der Deutschen Nationalbibliografie; detaillierte bibliografische Daten sind im Internet über http://dnb.d-nb.de abrufbar.
Alle in diesem Buch genannten Marken und Produktnamen unterliegen warenzeichen-, marken- oder patentrechtlichem Schutz bzw. sind Warenzeichen oder eingetragene Warenzeichen der jeweiligen Inhaber. Die Wiedergabe von Marken, Produktnamen, Gebrauchsnamen, Handelsnamen, Warenbezeichnungen u.s.w. in diesem Werk berechtigt auch ohne besondere Kennzeichnung nicht zu der Annahme, dass solche Namen im Sinne der Warenzeichen- und Markenschutzgesetzgebung als frei zu betrachten wären und daher von jedermann benutzt werden dürften.

Bibliographic information published by the Deutsche Nationalbibliothek: The Deutsche Nationalbibliothek lists this publication in the Deutsche Nationalbibliografie; detailed bibliographic data are available in the Internet at http://dnb.d-nb.de.
Any brand names and product names mentioned in this book are subject to trademark, brand or patent protection and are trademarks or registered trademarks of their respective holders. The use of brand names, product names, common names, trade names, product descriptions etc. even without a particular marking in this work is in no way to be construed to mean that such names may be regarded as unrestricted in respect of trademark and brand protection legislation and could thus be used by anyone.

Coverbild / Cover image: www.ingimage.com

Verlag / Publisher:
LAP LAMBERT Academic Publishing
ist ein Imprint der / is a trademark of
OmniScriptum GmbH & Co. KG
Heinrich-Böcking-Str. 6-8, 66121 Saarbrücken, Deutschland / Germany
Email: info@lap-publishing.com

Herstellung: siehe letzte Seite /
Printed at: see last page
ISBN: 978-3-659-64313-2

Copyright © 2015 OmniScriptum GmbH & Co. KG
Alle Rechte vorbehalten. / All rights reserved. Saarbrücken 2015

Reactive Power Planning

Author:
Beatriz Magalhes

Supervisor:
Vladimiro Miranda (Full
Professor)

January 2015

"The limits of the possible can only be defined by going beyond them into the impossible."

Sir Arthur C. Clarke

Abstract

Currently, there is an increasing demand from the transmission systems efficiency, due to the difficulties in their further expansion imposed by several policies. The major requirement from these systems is a high transmission capacity combined with a voltage profile within the ranges at which customers operate. The presence of wind based power penetration in these systems enhances the existing challenges due to its fluctuating nature.

Reactive Power Planning is one of the most intricate power systems optimization problems. It is defined as the optimal location of reactive power compensation devices as well as determining their types and sizes while minimizing the investment and power losses costs and maintaining an adequate voltage profile.

This thesis presents an application of the Differential Evolution Particle Swarm Optimization algorithm for solving the Reactive Power Planning problem with wind power penetration. Fixed and switched capacitor banks, TCR, SVC and STATCOM are installed and sized to minimize the investment and power losses costs and the voltage deviations from the nominal value. Besides a deterministic model, where only one scenario exists, a probabilistic model is implemented. This approach shows the reactive power compensation on a more realistic network that has load and wind generation scenarios with associated probabilities of occurrence.

Results are presented for a modified IEEE 118-bus System. Initially, four different deterministic scenarios are tested and the results demonstrate that planning the purchase and installation of new devices in the network based solely on this kind of scenarios is not reliable. Then, two different values of voltage deviation validate the probabilistic model and offer insight on its behaviour. The placement of the five kind of devices as well as the optimization of the transformer taps settings and the voltage on PV nodes contribute for the reduction of losses costs and maintaining a good voltage profile, while minimizing investment costs. Finally, it is presented a trade-off analysis between the maximum allowed voltage deviation and the power losses.

Keywords: Differential Evolution Particle Swarm Optimization; FACTS devices; Reactive Power Planning; Wind power integration.

Acknowledgements

First of all I would like to acknowledge my thesis supervisor, Professor Vladimiro Miranda, for having arranged for me such an interesting theme and for the valuable insights and guidance that improved this work. I would also like to acknowledge Leonel Carvalho for the constant availability and for all the help and support. Without all the help, motivation and passion from both this would be a poorer work definitely. I acknowledge Carolina Marcelino as well for all the support and encouragement during these five months.

The rest of the acknowledgements are for those who made me who I am today and therefore are made in my mother tongue.

Aos meus pais por sempre me terem incentivado e apoiado, em particular na minha vida académica e por fazerem de mim um poço de trivia inútil e à mãe pelas horas dispensadas para corrigir o inglês da minha tese. Ao meu irmão Raúl, por ao mesmo tempo ser um companheiro e alguém para quem quero sempre ser um bom exemplo. Aos meus avós pelo apoio constante e pela segunda casa. Às minhas tias pela amizade e por serem, em conjunto com a avó, segundas mães. Ao Titó por sempre me ter incentivado a saber mais.

Ao João, por ser uma pessoa tão importante na minha vida, por estar sempre ao meu lado, por me acalmar nos momentos de pânico da tese, pelo esforço por compreender este trabalho apesar de ser fora da sua zona de conforto e por me motivar a constantemente aprender algo novo.

Finalmente, a todos os elementos da Tuna Feminina de Engenharia, que nela integraram durante estes cinco anos e fizeram deles os cinco anos mais incríveis da minha vida. À Catarina, Inês Vigário, Tatiana, Filipa, Ana Carraca e Inês Machado pelo companheirismo e todos os momentos destes cinco anos; à Catarina pela companheira constante, à Tatiana e Catarina pelas horas infindáveis de tricot e à Inês pelo incentivo para não deixar cadeiras e pela playlist altamente e ideal para a tese. À Sónia, à Maria João e à Sofia pelos excelentes exemplos, pelos conselhos e pela amizade. À Inês Cunha e à Ângela pelas épocas de exames altamente. À Madalena e à Ana Faria pelas dores de cabeça e pela cromice. E à Isabel (membro honorário) pela surpresa que proporcionou e por contribuir para que a minha queima de finalista ainda fosse melhor.

Beatriz Magalhães

Contents

List of Figures

List of Tables

Abbreviations

LAH	List Abbreviations Here
AC	Alternating Current
ANOVA	ANalysis Of VAriance
DE	Differential Evolution
DEEPSO	Differential Evolutionary Particle Swarm Optimization
EA	Evolutionary Algorithms
EP	Evolutionary Programming
EPSO	Evolutionary Particle Swarm Optimization
FACTS	Flexible AC Transmission Systems
FC	Fuzzy Clustering
GA	Genetic Algorithm
PSO	Particle Swarm Optimization
QEA	Quantum Evolutionary Algorithm
RPP	Reactive Power Planning
SA	Simulated Annealing
SVC	Static VAR Compensator
STATCOM	STATic Synchronous COMpensator
TCR	Thyristor Controlled Reactor
VAR	Volt-Ampere Reactive

Chapter 1

Introduction

1.1 Context and Motivations

Nowadays, electric energy has an extreme importance in the highly industrialized societies and it is indispensable in both homes and industry causing an increasing consumption as well as higher quality standards. Thereby, it has become necessary to efficiently manage the existing networks, being very selective and well thought the investments for the creation of new infrastructures.

Traditionally, the electric sector was based in hydroelectric, thermal and nuclear plants. Although hydroelectric plants are clean and efficient, the locations for its installation are becoming scarce. On the other hand, the fossil fuels necessary for the thermal plants are running out, generate a high dependency on their producing countries and are some of the main causes of the CO_2 emissions. This has caused a major change in the electric systems paradigm with governments implementing goals and objectives aimed at the extensive integration of energy from renewable sources in the electric networks.

From all the renewable technologies, wind power stands out since it has had a significant increase of integration in the networks. As it is clear by Figure 1.1, the wind power installed worldwide has been increasing, with a growth of 63.7% since 2010.

1

FIGURE 1.1: Evolution of wind power installed in the world [1].

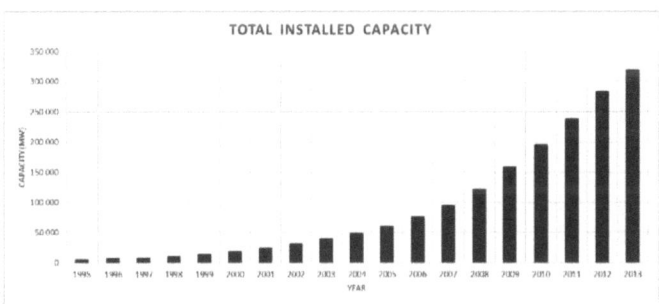

FIGURE 1.1: Evolution of wind power installed in the world [1].

Despite its advantages, wind power causes technical problems due to the difficulty in controlling its voltage and reactive power, the fact that it is not dispatchable and its fluctuating nature.

The Reactive Power Planning problem is defined as the optimal allocation of reactive power compensation devices as well as determining their types and sizes while minimizing the investment and power losses costs and maintaining a voltage profile adequate to the consumers.

The most common solutions for reactive power compensation are the capacitor banks because they are relatively cheap when considering their compensation capacity.

Flexible Alternating Current Transmission Systems (FACTS) devices control the power flow in the system, allow a better use of the transmission line capability and improve the system's security and stability. Regarding the reactive power compensation, there are FACTS that have great advantages: TCR, SVC and STATCOM. The SVC and TCR are based on sets of reactive elements controlled by power electronic devices (mainly thyristors) and which allow to dynamically adjust the reactive power injected in the network. The STATCOM operation rests in power electronic devices with forced switching, allowing an independent control of the voltage amplitude and phase. Although the FACTS devices possess many benefits, they have high costs. Therefore, it is not easy to obtain a cost-effective utilisation.

1.2   Objectives

This research work emerged with the main purpose of proposing an application to solve the RPP problem in systems with wind power penetration. The method decides the installation of hybrid capacitor banks, TCR, SVC and STATCOM to minimize the investment and losses costs and the voltage deviations from the nominal value. Further developments were made to account for the uncertainty in the system, both in wind generation and load. The multiple scenarios increase the complexity of the problem which further justifies the use of a recent and powerful meta-heuristic algorithm: DEEPSO.

The objectives of this thesis are:

- Implementation and validation of a RPP solving application with DEEPSO that accounts for systems with wind penetration;

- Evaluation of the viability of performing the planning of the reactive power compensation based on deterministic models;

- Study of the results provided by the probabilistic method;

- Analysis of the conflicting objectives behaviour.

1.3   Document Structure

This document has five chapters, including this introduction. In Chapter 2, the existing technologies for reactive power planning are presented as well as the state of the art of the field.

In Chapter 3, a mathematical formulation of the problem is presented along with the models necessary for its implementation. Then, the DEEPSO algorithm is described as well as the modeling of the problem so that it can be solved with this algorithm. Finally, the statistical tools for the parameters tuning are explained.

Chapter 4 shows the results obtained during the course of the parameters tuning process. Then, the application results are analysed, the ones from the deterministic model and from the probabilistic model.

Chapter 5 offers the conclusions of the developed work as well as references to future works that may be developed having this thesis as a starting point.

Chapter 2

State of the Art

2.1 Transmission networks

Originally, the electric sector had a vertically integrated structure, meaning that the same company would own the production, transmission and distribution of energy. Additionally, each company had the monopoly of its area system and, therefore, competition was inexistent. This structure combined with the economic environment at that time resulted on a very low degree of uncertainty and risk whereby the planning tasks were far less complex and there was an oversizing of the installed equipment, since at one point in the future they would be justifiable.[3]

During the eighties and nineties the sector went through a restructuring process, which resulted in the separation of the vertically integrated companies. Consequently, generation, transmission and distribution became separate activities. Whereas the generation segment became a very competitive one, the transmission networks remained under monopoly, regulated by public authorities.

Nowadays there is great concern in reducing the emissions which translate in incentives to use endogenous resources, especially wind and hydroelectric. Wind energy penetration continually supplying significant amounts of energy in some regions. [4] However, the high penetration levels pose a challenge to the electric system due to the fluctuating nature of the wind.

In conclusion, due to the high levels of production and consumption of energy, it is required a high capacity to the transmission systems, while keeping the voltage at decent levels for all consumers. The most obvious solution would be to create more lines but, due to several issues and policies, that is not possible. Thus, it is necessary to maximize the energy transmission efficiency so that a maximum amount of power is transferred with low levels of losses.

2.2 Reactive Power Compensation

Reactive power compensation is an important matter in power systems, thus there is a significant number of ways to accomplish it, either by connecting new devices to the network or by optimizing the status of the existing ones.

2.2.1 Dynamic and fixed capacitance devices

These devices include typically transformer taps, which is a connection in a transformer winding which allows the selection of a certain number of turns. The result is a variable turns ratio that allows stepped voltage regulation of the output. The synchronous generator is also included in this kind of devices but won't be used in this dissertation because, as presented below in section 2.2.2, the STATCOM is a device with the same functions but more benefits comparatively to the synchronous generator. The devices mentioned above are dynamic.

Capacitors banks are the fixed capacitance devices. They consist on capacitors that are connected in series and/or parallel and it reactance output depends of these connections. There are also switched capacitor banks which reactance output has a minimum fixed value and can be increased by fixed levels.

2.2.2 Flexible Alternating Current Transmission Systems (FACTS)

According to [5], FACTS are alternating current transmission systems incorporating power electronic-based and other static controllers to enhance the controllability and to increase power transfer capability. The main purposes of FACTS devices are to control the power flow in the system, better use of the transmission lines capability according to its thermal limits and to improve the system security and stability.

FACTS devices have various functions and can play distinct roles in the power system, which are presented in Table 2.1. Since we are only aiming to reactive power planning, the FACTS used in this dissertation were the Thyristor Controlled Reactor (TCR), the Static VAR Compensator (SVC) and the Static Synchronous Compensator (STATCOM). These will be presented in some detail.

TABLE 2.1: Roles of FACTS devices [2]

Operating problem	Corrective action	FACTS controller
Voltage limits		
Low voltage at heavy load	Supply reactive power	STATCOM, SVC
High voltage at low load	Absorb reactive power	STATCOM, SVC, TCR
High voltage following an outtage	Absorb reactive power; prevent overload	STATCOM, SVC, TCR
Low voltage following an outage	Supply reactive power; prevent overload	STATCOM, SVC
Thermal limits		
Transmission circuit overload	Reduce overload	TCSC, SSSC, UPFC, IPC, PS
Tripping of parallel circuits	Limit circuit loading	TCSC, SSSC, UPFC, IPC, PS
Loop flows		
Parallel line load sharing	Adjust series reactance	IPC, SSSC, UPFC, TCSC, PS
Postfault power flow sharing	Rearrange network or use thermal limit actions	IPC, TCSC, SSSC, UPFC, PS
Power flow direction reversal	Adjust phase angle	IPC, SSSC, UPFC, PS

TCR is composed of an anti-parallel thyristor pair, each conducting on alternate half cycles, in series with a fixed inductance. In practice, it acts as a controllable susceptance that varies in a continuous manner. Physically, the SVC is a TCR in parallel with a capacitor bank. Its function is voltage regulation by suitable control of its reactance, generating or absorbing reactive power. Figure 2.1 shows the reactance models of these two devices.

The STATCOM is analogous to a synchronous compensator, without the moving parts, with faster response to control actions. It outputs capacitive or inductive current independently of the system's voltage.

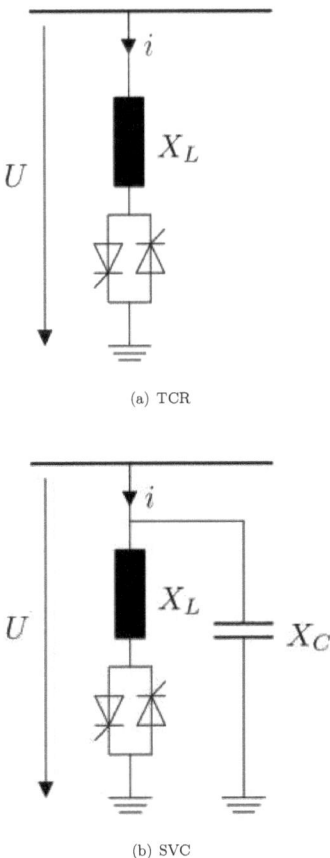

(a) TCR

(b) SVC

FIGURE 2.1: Reactance models.

2.3 Reactive Power Planning (RPP)

Reactive power causes losses at generation and transmission in the power systems. The production and absorption of this energy also affects the system voltage profile, which is important to be kept at adequate levels since it has effects on the system security and its ability to prevent voltage collapse. However, the carriage of reactive power through the transmission system, even when the objective is its compensation, incurs in power losses which is why a careful and well thought reactive power planning is extremely important.

Although there are different configurations and purposes for reactive power compensation devices, different objectives can be achieved through their placement and sizing. The parameters considered in the objective functions of the RPP problems are: power loss reduction, installation and operation cost minimization and voltage profile improvement. The reactive power compensation methods are various, from the settings of the transformers taps and the voltage on the regulated nodes to devices inserted in the network, which include capacitor banks and FACTS, and also the reactive power injection from wind farms.

The objective functions and constraints present in the RPP problems are of an intricate complexity which is the reason why the algorithms used to solve it follow different methodologies. The majority of the algorithms fall into three main categories: conventional mathematical methods like linear and nonlinear programming, intelligent searches such as evolutionary algorithms and simulated annealing, and artificial intelligence methods as fuzzy set and neural networks. [6]

There is an extensive number of publications concerning RPP, where the authors use different algorithms and have different objectives. The next paragraphs contain a brief description of the most significant work in the field. A table with the objective of standing out the main conclusions obtained is presented in the end of the chapter.

In the beginning, Linear Programming (LP) was used [7, 8] for both allocation and operation planning of reactive power. In [8] an algorithm based on the successive quadratic programming is used to achieve maximum active power transmission margin through reactive power equipment operation and planning, while satisfying system constraints.

In [9], Padhy et al. used the Genetic Algorithm (GA) to identify the optimal location and control parameters of FACTS with the purpose of minimizing the reactive power losses and generation. The algorithm was tested for the IEEE-30 system with multiple Thyristor-Controlled Series compensator and it has also been demonstrated that it can be applied to larger systems without further computational difficulties.

Miranda et al. solve the optimal capacitor placement and control with a hybrid mathematical GA, the Genetic Based Algorithm with Gradient Search. The main objectives were the minimization of investment costs, energy losses and voltage limits violation.

The hybrid GA is also compared with the Simple GA and the Genetic Based Algorithm. [10]

The strength of Evolutionary Algorithms (EA), specifically Evolutionary Programming (EP), for power system planning problems, namely RPP, is demonstrated in [11, 12]. Gopalakrishnan et al. used a hybrid EP algorithm in [12] with the objective of minimizing operation cost and the real power loss and improving voltage profile. First, the classical EP creates the cases on the base level and afterwards sequential quadratic programming locates the optimum.

Chen et al. [13] use GA to find optimal shunt capacitor location and control the bus voltage, achieving good results on the peak power system. In order to reach the defined goal of 1 p.u. voltage on bus, the fitness function is built with penalty factors.

Keko et al. [14] resort to the Evolutionary Particle Swarm Optimization (EPSO) to plan the installation of fixed and switched capacitor banks, as well as, transformer taps setting and voltage on regulated nodes on systems with multiple load levels and with different structural scenarios. The method is applied on two different systems, with various load levels and under contingency scenarios and a trade-off analysis between admissible voltage deviation and power losses is performed.

In [15], Eghbal and Araby present a comparative study between the use of GA, Particle Swarm Optimization (PSO) and EPSO to solve the RPP problem, by the installation of slow and fast VAR devices during normal and contingency states, with the implementation of a trade-off between security and economy. Both GA, PSO and EPSO led to a substantial saving in cost, even though the system security level is maintained, with EPSO having the best convergence, followed by PSO.

Chen and Ke [16] apply the two-layer Simulated Annealing (SA) to the multi-objective optimization problem. The optimal placing of VAR sources while minimizing losses and keeping constraints within its boundaries is the objective function applied. Apart from an IEEE test system, the algorithm is also applied to the Tai-power system.

Rajkumar and Devaraj [17] determine the reactive power control variables by means of the PSO algorithm. The created method is applied to the IEEE 30-bus system with effectiveness in maximizing the voltage profile and minimizing the operation and reactive power allocation cost. The candidate bus to the capacitors location are the weak buses, identified by the minimum singular value method.

Mao and Li [18] introduce voltage stability as an objective to the RPP problem. The optimal location and capacity of the new installed reactive power compensation is achieved resorting to a SA PSO hybrid together with a Fuzzy method. The voltage stability of the system is improved at the same time as the system losses are diminished and the voltage quality is enhanced.

Yiqin [19] uses an improved Tabu Search Algorithm to minimize the sum of the active power loss and the cost of reactive power equipment in the presence of different load levels. The algorithm is improved by modifying how the initial solutions are generated and how the search is conducted, showing a strong ability to escape from local optimal solutions. Optimal results to the problem are achieved, being a more representational one due to the multiple load levels.

Wang et al. [20] combine Fuzzy Clustering (FC) with learning automata to solve the VAR planning problem. FC selects the candidate nodes for new VAR sources and a P-model learning automata algorithm provides the multi-objective optimization results. Tested on the IEEE 57-bus, it is demonstrated that the learning automata is feasible to produce trade-off analysis between the generation cost, VAR cost, voltage stability and active power loss.

Hong and Pen [21] resort to the Quantum Evolutionary Algorithm to plan the transformer taps, existing wind generator voltages and static capacitors installation and to the Markov model for the wind power generations and bus loads. The objective function of the QEA features the real power loss and static capacitor cost minimization as well as satisfaction of the constraints. The model is simulated in two different systems and the results are compared to the ones obtained from the traditional GA.

Differential Evolution (DE) is used in [22] to allocate new reactive power sources and optimize the ones available in the system. Furthermore, DE is compared with another evolutionary technique, EP.

Rahmani et al. [23] conjugate the RPP problem with short-term transmission network expansion with the utilization with the Real Genetic Algorithm, identifying weak buses for new reactive power sources allocation and include constraints deviations and cost of the configuration on the fitness function. The method is validated by its application to a well-known system. The authors point out that the location of reactive power sources at load buses elevates the transmission capacity and drop the investment cost.

Kanokbannakorn and Audomvongseree [24] also combine the RPP problem with another power systems problem. In this case, the aim is to enhance the system reliability through the Monte Carlo simulation and plan the reactive power compensation through the optimal placing and sizing of capacitors by analysis of the power loss sensitivity index. The problem's objective function is also more fit with the expected reduction of the interruption cost.

Alonso et al. [25] found the optimal location for different wind farms and SVC units through GA, which handles both the multi objective goals and the variable wind power production in different scenarios. The validation of the algorithm is done by its application to a 140-bus network, where several solutions are obtained by the variation of the priority factors and weights.

An application is presented in [26] that provides reactive power compensation through SVC in systems with wind power penetration. Niu and Xu use the Monte Carlo simulation to account for the uncertainties inherent to the variable wind speed and the DE algorithm places SVC with the objective of improvement of the voltage profile and minimization of real power losses. There is also present in the objective function a penalty to ensure a high voltage quality at the wind farms connection point. The effectiveness of this application in providing satisfactory results is demonstrated with a modified IEEE 30-bus system with 6 scenarios of wind speed.

In the past few years, there was a large amount of research conducted in the RPP subject. Even though the algorithms used to solve the problem have different natures, there is a trend in using meta-heuristics. Since its features make them favourable to apply to multi-objective problems with a large number of variables, these kind of algorithms are the more appropriate to optimize the location and size of the reactive power compensation devices in power systems. In this thesis, a recent algorithm will be applied to this problem and its results analysed in detail.

2.4 Conclusions

After this literary review, it is safe to say that there is an absence of work that solves the RPP problem with the inclusion of a wide range of FACTS devices. There is also no documentation that combines all reactive power compensation devices, including hybrid ones, with uncertainty in the generation and load. This is the gap that the present work pretends to fill, creating a realistic model with the application of the brand new DEEPSO algorithm to its resolution.

TABLE 2.2: Literature review summary.

Authors	Objectives	Control variables	Algorithm
Gopalakrishnan et al. [12]	Minimize operation cost, real power loss and new reactive power sources cost and improve voltage profile.	Reactive power sources location and size	Hybrid Evolutionary Programming
Keko et al. [14]	Minimize active power losses, capacitor investment cost and voltage limits violation.	Transformer taps settings, voltage on voltage-regulated nodes and amount of reactive power and location of new capacitor banks.	Evolutionary Particle Swarm Optimization
Chen and Ke [16]	Minimize real power losses, investment costs, voltage deviation and constraints violation.	Reactive sources location and size.	Simulated Annealing
Rajkumar and Devaraj [17]	Maximize voltage profile and minimize operation and reactive power allocation cost.	Generators voltage, transformer taps settings and capacitor size.	Particle Swarm Optimization
Yiqin [19]	Minimize active power losses and reactive power equipment cost.	Generators voltage, transformer taps settings and reactive power compensation location and size.	Tabu Search Algorithm
Wang et al. [20]	Minimize generation and VAR cost, enhance voltage stability and minimize active power loss.	New reactive power sources installation	Fuzzy clustering with learning automata
Hong and Pen [21]	Minimize real power losses and static capacitor cost.	Transformer taps settings, wind generator voltages and static capacitors installation.	Quantum Evolutionary Algorithm
Cuello-Reyna and Cedeno-Maldonado [22]	Minimize real power losses and new reactive power sources.	Reactive power sources location and size.	Differential Evolution
Alonso et al. [25]	Minimize real power losses, maximize voltage loadability and minimize constraints violation.	SVC and wind farms location and size	Genetic Algorithm

Chapter 3

Problem Modelling

The Reactive Power Planning is a nonlinear optimization problem involving multiple variables of different nature and constraints. Due to these reasons meta-heuristics are the most appropriate methods for its resolution, as shown in Chapter 2.

The main objective of this chapter is to allow the reader to understand all the steps that were taken in order to solve the proposed problem as well as the necessary tools to accomplish this. Initially a mathematical formulation of the problem is presented along with the models necessary for its implementation. The DEEPSO algorithm is then described followed by the problem modelling to solve it with this algorithm. Finally, the statistical tools for the parameters tuning are presented.

3.1 Problem formulation

The objective function of the problem has the main objective of minimizing the active power losses and investment cost. It can be written as

$$\min C(Q_C) + \sum_{j=1}^{N_S} p(S) C(P_L)_j, \qquad (3.1)$$

where $C(Q_C)$ represents the cost of newly installed compensation devices, $C(P_L)_j$ the losses cost on scenario j, $p(S)$ the probability of scenario S and N_S the total number of scenarios. The investment cost accounts for the devices purchased and installed in the network

$$C(Q_C) = Q_{C,FCB} \cdot C_{FCB} + Q_{C,SCB} \cdot C_{SCB} + \sum_{i=1}^{N_C} b_{TCR,i} \cdot C_{TCR} +$$
$$+ \sum_{i=1}^{N_C} b_{SVC,i} \cdot C_{SVC} + \sum_{i=1}^{N_C,PQ} b_{STATCOM,i} \cdot C_{STATCOM},$$

(3.2)

where $Q_{I,FCB}$ and $Q_{I,SCB}$ are the capacitor banks levels installed, $b_{TCR,i}$, $b_{SVC,i}$ and $b_{STATCOM,i}$ are the FACTS binary variables, which have value one if a FACTS device is installed in node i. C_{FCB}, C_{SCB}, C_{TCR}, C_{SVC} and $C_{STATCOM}$ are the costs of the devices. N_C is the total number of candidate nodes and $N_{C,PQ}$ is a subset of the candidate nodes, the ones that are PQ and therefore where STATCOM can be installed.

The losses cost calculation is made for each scenario j according to the equation

$$C(P_L)_j = P_{L_j} \cdot MC_j \cdot d_j$$

(3.3)

where P_{L_j} are the total active power losses in MW, MC_j is the energy marginal cost in €/(MW.h) and d_j is the duration of the scenario in hours.

The constraints of the problem are divided in power flow equalities, bus voltage limits and reactive power compensation devices limits. The power flow equalities consist of the active power balance for each node i

$$[P_{gi}] - [P_{di}] - [P_i] = 0,$$

(3.4)

where $[P_{gi}]$ is the active power generation vector, $[P_{di}]$ the load active power vector and $[P_i]$ the active power injected in the node; and the reactive power balance

$$[Q_{gi}] + [Q_{ci}] - [Q_{di}] - [Q_i] = 0,$$

(3.5)

where $[Q_{gi}]$ is the reactive power generation vector, $[Q_{ci}]$ the compensation devices reactive power vector, $[Q_{di}]$ the load reactive power vector and $[Q_i]$ the reactive power injected in the node.

The bus voltage limits are

$$V_i^{min} \leq V_i \leq V_i^{max},$$

(3.6)

where V_i represents the voltage on bus i.

The reactive compensation devices limits consist on the TCR and SVC firing angle, α, limits

$$\alpha^{min}_{TCR,i} \leq \alpha_{TCR,i} \leq \alpha^{max}_{TCR,i}, \tag{3.7}$$

$$\alpha^{min}_{SVC,i} \leq \alpha_{SVC,i} \leq \alpha^{max}_{SVC,i}, \tag{3.8}$$

the STATCOM voltage, ΔV, and its reactive power generation, $Q_{g,STATCOM}$, limits

$$\Delta V^{min}_i \leq \Delta V_i \leq \Delta V^{max}_i, \tag{3.9}$$

$$Q^{min}_{g,STATCOM} \leq Q_{g,STATCOM} \leq Q^{max}_{g,STATCOM}, \tag{3.10}$$

the transformer taps settings, T_m, limits

$$T^{min}_m \leq T_m \leq T^{max}_m, \tag{3.11}$$

the capacitor banks reactive power, Q_C, on each node i

$$Q^{min}_C \leq Q_C \leq Q^{max}_C, \tag{3.12}$$

and the FACTS binary variables, b, limits

$$b^{min}_{TCR,i} \leq b_{TCR,i} \leq b^{max}_{TCR,i}, \tag{3.13}$$

$$b^{min}_{SVC,i} \leq b_{SVC,i} \leq b^{max}_{SVC,i}, \tag{3.14}$$

$$b^{min}_{STATCOM,i} \leq b_{STATCOM,i} \leq b^{max}_{STATCOM,i}. \tag{3.15}$$

3.2   Models for power flow studies

3.2.1   TCR and SVC

As presented in 2.2.2, the TCR is composed by an inductance in series with a thyristor switch, while the SVC is a TCR in parallel with a capacitor. Thereby, the modelling of both devices is very similar, which is why it is presented in the same section.

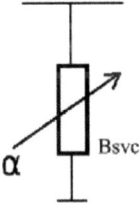

FIGURE 3.1: SVC firing angle model.

Figure 3.1 presents the firing angle model for the SVC and the TCR. This is the model which will be used the power flow studies with these device, where the equivalent reactance is a function of the firing angle α. The TCR reactance B_{TCR} is given by

$$B_{TCR} = \frac{2(\pi - \alpha_{TCR}) + \sin(2\alpha_{TCR})}{\pi X_{L,TCR}}, \qquad (3.16)$$

where α is the firing angle and X_L is the total inductance. The expression for the SVC reactance

$$B_{SVC} = \frac{\pi X_{L,SVC} - X_C[2(\pi - \alpha_{SVC}) + \sin(2\alpha_{SVC})]}{\pi X_{L,SVC} X_C}, \qquad (3.17)$$

where α and X_L are the same from Equation (3.16) and X_C is the total capacitance.

The firing angles ranges from $\pi/2$ (maximum reactance) to π (minimum reactance). In Figure 3.2 we can observe the TCR and SVC's reactance variation with the firing angle. As it can be perceived, the equivalent reactances do not present any discontinuities in the range used.

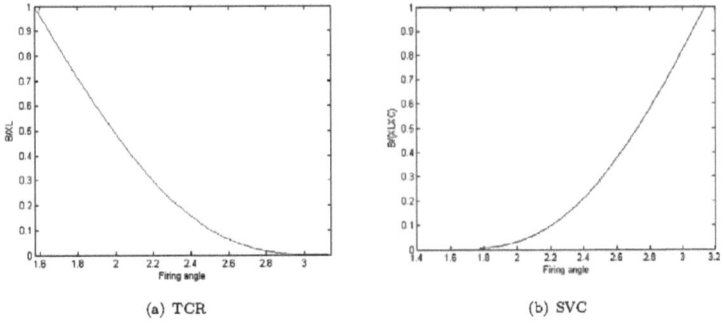

(a) TCR   (b) SVC

FIGURE 3.2: Reactance in order to the firing angle.

3.2.2 STATCOM

As also presented in 2.2.2, the STATCOM supplies a voltage ΔV in phase with the network voltage. According to the signal of this voltage, the system supplies or absorbs reactive power from the system. In power flow studies, the STATCOM is considered a synchronous generator due to its modelling similarities to this device. Firstly, the bus where the STATCOM is placed, which has to be PQ, becomes PV. Then, a synchronous generator with null real power output and voltage set to 1 pu is there connected. Last, the voltage on the now voltage-regulated node is set to the STATCOM's voltage, ΔV. It is also necessary to keep the reactive power generated by the STATCOM between its limits. As it will be presented ahead, this control will be made in the form of penalties.

3.3 Uncertainty modelling

3.3.1 Load model

The load variation in a time period is illustrated in a chart, called load profile. For the purpose of this work, a load probability profile was made with the three load scenarios, off-peak, normal and peak, presented in Figure 3.3. Each scenario has a probability, also present in the graph, where medium load is more probable than low and high load.

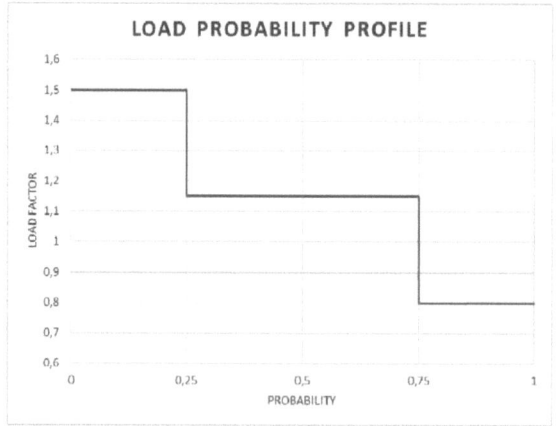

FIGURE 3.3: Load probability profile.

3.3.2 Generation model

The generation from wind depends on the location of the farm and varies with time due to the fluctuations in wind speed. Since the wind turbine output depends on the wind speed it is first necessary for it to be modelled according to its probabilistic behaviour.

One of the most common and accepted model for wind speed variation is the Weibull distribution. The two-parameter Weibull probability density function is given by

$$f(U) = \frac{\beta}{\alpha} \left(\frac{U}{\alpha} \right)^{\beta-1} e^{-\left(\frac{U}{\alpha}\right)^{\beta}}, \tag{3.18}$$

where U is the wind speed, α the scale parameter and β the shape parameter. The shape parameter ranges from 1 to 3 and small values of this parameter mean a high variability of wind speed. The scale parameter is proportional to the mean wind speed and it is a measure for the distribution's characteristic wind speed. The probability density function used in this work is presented in Figure 3.4, with a scale parameter of 7.5 m/s and shape 2.

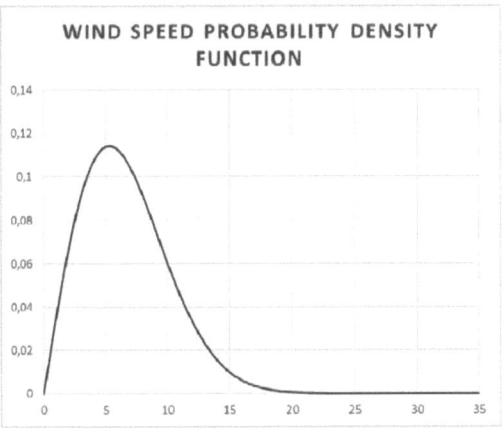

FIGURE 3.4: Weibull probability density function.

The Weibull cumulative distribution function allows the determination of the probability of certain ranges of wind speed. It is the integral of the probability density, thus given by

$$F(U) = 1 - e^{-\left(\frac{U}{\alpha}\right)^{\beta}}. \tag{3.19}$$

The cumulative distribution associated with the probability density function (3.18) is:

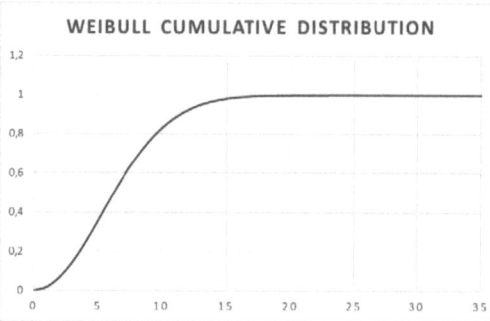

FIGURE 3.5: Weibull cumulative distribution.

The power output of a wind turbine varies according to the wind speed and the typical wind turbine output curve is presented in Figure 3.6. Wind turbines have three different operation modes. The cut-in and cut-out speeds correspond to the minimum and maximum speed values necessary for the turbine to output power in order to guarantee the conservation of the machine. The second operation mode is when the wind speed varies from the cut-in speed to the rated output speed where the turbine's power output is proportional to the cube of the wind velocity. At the rated output speed the power output reaches its maximum value (rated output power) and it is constant up to the cut-out speed, this being the third operation mode.

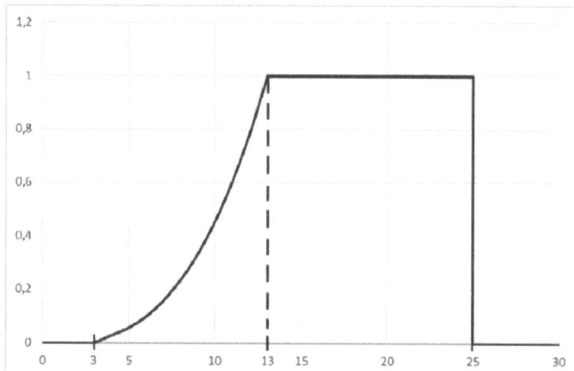

FIGURE 3.6: Wind turbine power output.

By overlapping the turbine power with the wind speed models one can obtain the probability distribution of the wind power. Defining the three wind generation scenarios as low, medium and high wind, the power for each is the average of the corresponding interval. Table 3.1 presents the three intervals' wind speed ranges, probability and power.

TABLE 3.1: Wind scenarios.

Velocity - U (m/s)	Prob.	Power - P(U) (MW)
[0;5] U]25; 30]	0.3588	5.121
]5; 13]	0.5916	84.879
]13; 25]	0.0496	180

For input on the existing system, the wind generation was modelled as a load on PV nodes. [25] The wind farms are generating active power therefore the sign of the active power is negative. On the other hand, the reactive power is being consumed by the farm, thus having positive sign, and is calculated based on the expression

$$Q = \tan \phi \cdot P, \tag{3.20}$$

where ϕ is the angle corresponding to a power factor of 0.95.

3.4 DEEPSO

The Differential Evolutionary Particle Swarm Optimization (DEEPSO) algorithm combines the concept of rough gradient from DE with the self-adapting particle movements of EPSO. This last one is by itself a hybrid that obtains a higher efficiency from PSO by combining it with the selection and self-adaptiveness of EA.

DEEPSO is a population based algorithm, with a multidimensional search space. On each iteration there is a set of particles each containing a possible solution to the problem as well as the strategic parameters. The set of particles compose the swarm. Every particle has a position, representing a candidate solution to the problem, which is produced every iteration according to

$$X_i^{new} = X_i + V_i^{new}, \tag{3.21}$$

where V_i^{new} is the velocity of the particle and is obtained by the expression

$$V_i^{new} = w_{i1}^* V_i + w_{i2}^*(X_{R1} - X_{R2}) + P[w_{i3}^*(b_G^* - X_i)] \qquad (3.22)$$

where X_{R1} and X_{R2} are two particles sampled from the population and * marks the parameters that undergo mutation.

The particle's movement depends on three factors: inertia, memory and cooperation. The inertia aims to maintain its movement on the same direction as previously. The memory causes the particle's movement attracted to the best position it has found so far, b_i. The cooperation term draws the particle to the global best found by the swarm in its past life, b_G.

The DEEPSO algorithm is a self-adaptive method. Hence, the weights that affect each term of the movement rule w_{ix} are mutated according to the equation

$$w_{ik}^* = w_{ik}^* + \tau N(0, 1) \qquad (3.23)$$

where τ is the mutation rate and $N(0, 1)$ is a random variable with Gaussian distribution. The mutation rate controls the mutations' amplitude and is fixed externally. [27]

On each iteration, the initial population is replicated, both populations have its weights mutated and reproduced according to its movement rule. Then, both populations are evaluated and the best fit is selected to continue the process, therefore there is direct competition between parent and son.

The information received by the particle from the swarm is disturbed in a way that the cooperation term may attract the particle to the neighbourhood of the global best. Equation (3.24) represents how this disturbance is created. In it, w_{i4} controls the disturbance's amplitude and is a normal weight, mutated and selected as all others.

$$b_G^* = b_G(1 + w_{i4}N(0, 1)) \qquad (3.24)$$

P is a diagonal matrix that controls the topology of the communication between the particles. It is a diagonal matrix which contains binary variables of value 1 with probability p and value 0 with probability $(1 - p)$, where p controls the passage of information in the swarm. [28] With this value, the communication between particles may vary between a star model, where all particles are aware of the global optimum at all times, and the selfish model, where no communication exists in the swarm. The one adopted in DEEPSO is a stochastic method that remains between the two extremes and it is demonstrated in [28] that a fine tuning of the communication probability leads to better results.

The DEEPSO procedure is based on the EPSO. The DE component is present in the memory term and, for minimization, the two particles X_{R1} and X_{R2} should be ordered according to their fitness value

$$
\begin{cases}
V_i^{new} = w_{i1}^* V_i + w_{i2}^*(X_{R1} - X_{R2}) + P[w_{i3}^*(b_G^* - X_i)] & \text{if } f(X_{R1}) < f(X_{R2}) \\
V_i^{new} = w_{i1}^* V_i + w_{i2}^*(X_{R2} - X_{R1}) + P[w_{i3}^*(b_G^* - X_i)] & \text{if } f(X_{R1}) > f(X_{R2})
\end{cases}
\tag{3.25}
$$

In the DEEPSO model, X_{R2} is equal to X so that only X_{R1} is sampled. X_{R1} may be sampled from the set of particles from the same generation, P_c, or from the set of past best particles, P_b. The number of times the particles are sampled also varies, either it is once in the beginning or it is sampled for each component of V, and it is calculated from an uniform recombination of the particles from the sampling set. These differences in the sampling of X_{R1} create four different versions of the DEEPSO algorithm:

- DEEPSO Sg: X_{R1} is sampled once from P_c and the movement rule is as follows:

$$
V_i^{new} = w_{i1}^* V_i + w_{i2}^*(X_{R1} - X_i) + P[w_{i3}^*(b_G^* - X_i)]
\tag{3.26}
$$

- DEEPSO Sg-rnd: Similar to DEEPSO Sg, but X_{R1} is sampled for each component of V, being formed from an uniform recombination of all the particles from P_c. On this variant, the size of memory of the vector that has the sampling set has the same size as the population.

- DEEPSO Pb: X_{R1} is sampled once from P_b. This variant's movement rule is:

$$
V_i^{new} = w_{i1}^* V_i + w_{i2}^*(b_{R1} - X_i) + P[w_{i3}^*(b_G^* - X_i)]
\tag{3.27}
$$

- DEEPSO Pb-rnd: the same as DEEPSO Pb, but b_{R1} is sampled for each component of V. The sampling set memory vector has an adequate size to save all the historical past best particles.

As presented in [29], the DEEPSO Pb-rnd is the one with best results, including when applied to power systems problems. Therefore, this is the DEEPSO variation used throughout the entire work.

3.5 Solving the RPP problem with DEEPSO

The method that was developed in this thesis solves the Reactive Power Planning problem resorting to the DEEPSO algorithm. However, it may be adapted to any other meta-heuristic without much effort. Another versatility of this model is that any other FACTS device may be used, even though the most relevant for reactive power compensation are already included. MATLAB was the environment in which the application was developed.

Two models were implemented. In the first, the deterministic, every load and generation are known. On the other hand, the probabilistic model accounts for the uncertainty in the load and in the generation from wind farms. Therefore, each combination of load and generation scenarios results in a single scenario. All the scenarios combined compose a time period.

Despite the differences between the methods, their implementation is quite similar. Therefore, firstly the main method will be presented, including the particle's composition, the main outline of the method and the fitness function. Then, the details of each model as well as their implementation will be discussed.

The particle's composition and the main outline of the procedure are presented followed by a more detailed explanation of the steps of the algorithm implementation.

The encoding of the particle is the following: firstly there are the FACTS binary variables; then, there are the TCR and SVC firing angles, the STATCOM voltage, the voltage-regulated nodes voltage, the transformer taps settings and the capacitor banks reactive power output. The FACTS binary variables appear once at the beginning of the particle and the control variables for the reactive compensation devices are repeated the same number of times as the number of scenarios.

With the described representation, a particle from our problem looks like the following:

$$X = [b_{TCR,1}, ..., b_{TCR,N_C}, b_{SVC,1}, ..., b_{SVC,N_C}, b_{STATCOM,1}, ..., b_{STATCOM,N_{C,PQ}},$$
$$\alpha^1_{TCR,1}, ..., \alpha^1_{TCR,N_C}, \alpha^1_{SVC,1}, ..., \alpha^1_{SVC,N_C}, \Delta V^1_1, ..., \Delta V^1_{N_{C,PQ}}, V^1_1, ..., V^1_{N_P V}, T^1_1, ..., T^1_{N_T},$$
$$Q^1_{C,1}, ..., Q^1_{C,N_C},$$
$$...$$
$$\alpha^{N_S}_{TCR,1}, ..., \alpha^{N_S}_{TCR,N_C}, \alpha^{N_S}_{SVC,1}, ..., \alpha^{N_S}_{SVC,N_C}, \Delta V^{N_S}_1, ..., \Delta V^{N_S}_{N_{C,PQ}}, V^{N_S}_1, ..., V^{N_S}_{N_P V}, T^{N_S}_1, ..., T^{N_S}_{N_T},$$
$$Q^{N_S}_{C,1}, ..., Q^{N_S}_{C,N_C}]$$

The control variables that compose the particle are divided in two types: discrete and continuous. The discrete variables are the FACTS binary variables, the transformer taps settings and the capacitor banks size. The last two types of variables vary in levels due to its physical features. All the other variables are continuous. Also, in the variables corresponding to capacitor banks, a reactance of zero means that there is no capacitor bank installed in that node. This approach could not be used with the FACTS control variables due to its continuous nature and therefore the probability of them having value zero would be much inferior to the probability of them having a value different than zero. The FACTS binary variables were created to solve this problem.

The procedure followed in the resolution of our problem is described in the flowchart of Figure 3.7. All the steps will now be explained as well as the differences between the models, when present.

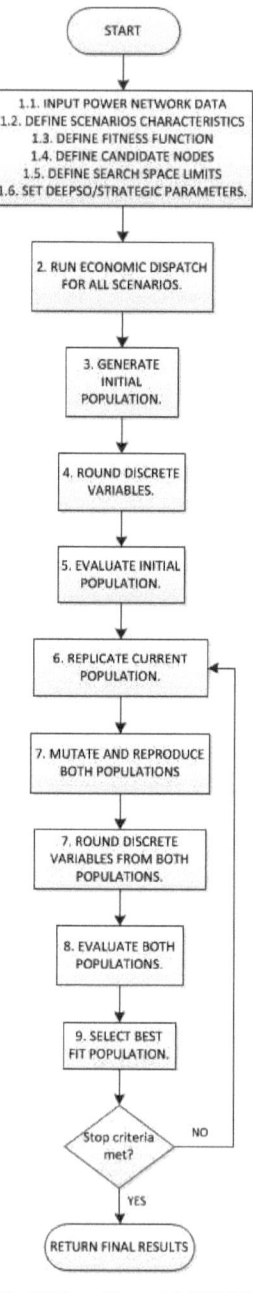

FIGURE 3.7: Solving the RPP problem with DEEPSO - main procedure.

Globally, the first step consists in the initialization of all the variables and parameters to our problem. In step 1.2, the scenarios characteristics consist in the load and generation for all the defined scenarios. In the deterministic model, only the load and wind generation are defined, while in the probabilistic model also requires the definition of the probability of each one of the scenarios. Since the load and wind generation are independent, each scenario probability is the generation probability times the load probability. These values are saved separately and when using each scenario, this simple calculation is made.

The fitness function parameters defined in 1.3 consist on the scaling factors, the purchase and installation cost of each type of devices and the limits for penalties. When the fitness function is explained in detail these will become clear.

The candidate nodes (step 1.4) are the nodes where capacitor banks, TCR and SVC may be installed. The candidate nodes for STATCOM are a subset from these, including only nodes that are PQ.

The search space limits consist on a part of the problem restrictions, which are forced and, therefore, are called hard constraints. The constraints included in the search space limits are the reactive power compensation control variables limits, Equations (3.7), (3.8), (3.9), (3.11) and (3.12), the FACTS binary variables limits, Equations (3.13), (3.14) and (3.15), and the voltage limits on the voltage regulated nodes

$$V_{i,PV}^{min} \leq V_{i,PV} \leq V_{i,PV}^{max}. \tag{3.28}$$

From the search space limits, the FACTS binary variables are the ones that need to be further explained. These variables are, as their denomination, binary. If they have value one, it means that a device of that type will be installed on the correspondent node. In the probabilistic model, where several scenarios are present, a device is installed for all scenarios. However, it may be decided not to use the device in one or more scenarios, by setting its control variable value to zero.

The second step, although separated from the initialization phase for its methodological differences, may still be considered a part of it. Here, a classical economic power dispatch without losses and with generators limits is performed for all the scenarios. This way the amount of energy from each generator and energy's marginal cost are obtained for each of the scenarios.

The following steps of the procedure are inherent to the DEEPSO algorithm presented in 3.4. The only one that is not mentioned is step 4 (as well as 8), which is necessary due to the presence of discrete variables.

The evaluation of each candidate solution, in steps 5 and 9, is made by a fitness function which is always very specific to the problem that is being solved by DEEPSO. In this case, the main goals of the problem are the minimization of the investment and power losses costs, while maintaining the voltage inside the operational limits on all nodes and the generated reactive power by the STATCOM in its limits. Our fitness function values the minimization of these factors. In order to get to the final result of the fitness, the procedure presented in Figure 3.8 must be followed.

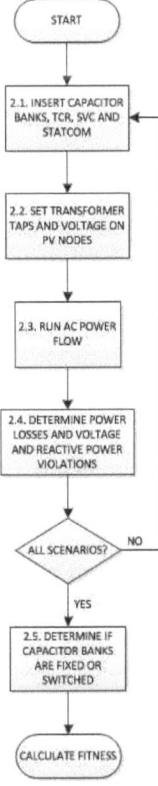

FIGURE 3.8: Solving the RPP with DEEPSO - fitness procedure.

On the first two steps, the base case is altered to match the present candidate solution. Thus the devices are installed in the correspondent nodes and the tap settings and the voltage on PV nodes are changed. Then, an AC power flow is performed on the modified network resorting to a MATLAB package, denominated MATPOWER. On step 2.4, with the power flow results, the power losses are calculated and the voltages and reactive power generation obtained. After going through all scenarios, the type of capacitor banks is determined following the criteria: if the size of the capacitor banks in a node is the same for all scenarios, a fixed capacitor bank is placed; otherwise, the capacitor bank will be switched, with a fixed reactance equal to the minimum from all the scenarios.This step is only present in the probabilistic method, since the deterministic has only one scenario present.

Finally, the fitness function is calculated according to the expression

$$\min fit = sfI \cdot C(Q_C) + \sum_{j=1}^{N_S} p(S) \times [sfL \cdot C(P_L)_j + sfV \cdot VP_j + sfQ \cdot Q_GP_j], \quad (3.29)$$

where $C(Q_C)$ and $C(P_L)$ are the investment and power losses cost from the problem formulation (3.1), VP is the voltage deviation penalty and Q_GP the STATCOM reactive power generation penalty. $p(S)$ is the probability of the scenario S and N_S the total number of scenarios. sfI, sfL, sfV and sfQ are the scaling factors associated to each objective and allow us to define the significance of each one in the global function. The presented fitness function is for the probabilistic method, since it accounts for the different scenarios with corresponding probabilities. The fitness function for the deterministic model is presented further ahead.

The constraints to which the fitness function is subject are the ones from the problem formulation, Equation (3.4) to (3.15).

The third and fourth part of the fitness deal with limits violations, in the form of penalties. These act as soft constraints leading the solution away from regions in the search space where the violations occur.

$$VP = \sum_{j=1}^{N_S} \left(\sum_{i=1}^{N_{PQ}} RV_{ij} + sfM \sum_{i=1}^{N_{PQ}} LV_{ij} \right) \quad (3.30)$$

$$Q_GP = \sum_{j=1}^{N_S} \left(\sum_{k=1}^{N_{STATCOM}} RQ_{ik} + \sum_{k=1}^{N_{STATCOM}} LQ_{ik} \right) \quad (3.31)$$

The penalties for both limits are obtained in the same way. Therefore, only the voltage penalty calculation is presented, for each scenario and each bus

$$LV = \begin{cases} V_{ij} - V_{N,i} & \text{if } V_{ij} < V_{min} \\ 0 & \text{if } V_{ij} \geq V_{min} \end{cases} \quad (3.32)$$

$$RV = \begin{cases} (V_{N,i} - V_{ij})^2 & \text{if } V_{ij} > V_{max} \\ 0 & \text{if } V_{ij} \leq V_{max} \end{cases} \quad (3.33)$$

The fitness function of the deterministic model is slightly different from (3.29), since only one scenario is present

$$Minfit = sfI \cdot IC_j + sfL \cdot LC_j + sfV \cdot VP_j + sfQ \cdot Q_G P_j. \qquad (3.34)$$

However, all the restrictions and components of the function are the same as the one explained, considering the number of scenarios, N_S, as one.

The stopping criteria of this algorithm is based on the maximum number of generations, and the number of unchanged generations. This is a variable that keeps track of the number of generations where there is no change in the fitness of the best solution. Using this variable, the algorithm is not forced to run a high amount of generations when the best fitness has already been achieved. However, an appropriate number for the maximum unchanged generations must be used so that the algorithm is not stopped on a local optimum solution.

3.6   Tools for parameters tuning

In order to determine the optimal values of significant factors to the algorithm used, we used the two-factor factorial design and the analysis of variance (ANOVA).

In the two-factor factorial design there are a levels for factor A and b factors B and each experiment is performed n times and it is usual to consider two levels for each factor, high and low.

The main effect of a factor is "the change in response produced by a change in the level of the factor". [30] However, when more than one factor are considered, an interaction between both factors may also be present, in other words, there may be different responses between the levels of one factor and the levels of the other. Figure 3.9 shows two graphs that intend to illustrate experiments with and without interaction between factors. When the interaction is high, the main effects may have small meaning or they can be masked. The only way to study interactions between variables along with its main effects is the factorial design, which is why it was used.

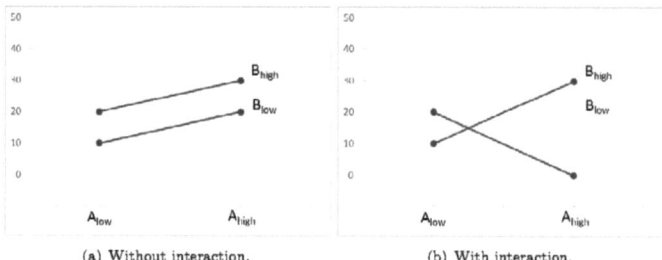

(a) Without interaction.   (b) With interaction.

FIGURE 3.9: Two experiments, with and without interaction.

A linear statistical model is used to describe the observations:

$$Y_{ijk} = \mu + \tau_i + \beta_j + \epsilon_{ijk}, (i = 1, 2, \ldots, a; j = 1, 2, \ldots b; k = 1, 2, \ldots, n) \qquad (3.35)$$

The data should be arranged as shown in Table 3.2, where y_{ijk} represents the observation for the ith value of A and the jth value of B for the kth replicate.

TABLE 3.2: Data arrangement for the two-factor factorial design.

		Factor B				Totals	Averages
		1	2	...	b		
Factor A	1	y_{111}, y_{112}, ...,y_{11n}	y_{121}, y_{122}, ...,y_{12n}		y_{1b1}, y_{1b2}, ...,y_{1bn}	$y_{1..}$	$\bar{y}_{1..}$
	2	y_{211}, y_{212}, ...,y_{21n}	y_{221}, y_{222}, ..., y_{22n}		y_{2b1}, y_{2b2}, ..., y_{2bn}	$y_{2..}$	$\bar{y}_{2..}$
	⋮						
	a	y_{a11}, y_{a12}, ...,y_{a1n}	y_{a21}, y_{a22}, ..., y_{a2n}		y_{ab1}, y_{ab2}, ..., y_{abn}	$y_{a..}$	$\bar{y}_{a..}$
Totals		$y_{.1.}$	$y_{.2.}$	...	$y_{.b.}$	$y_{...}$	
Averages		$\bar{y}_{.1.}$	$\bar{y}_{.2.}$	...	$\bar{y}_{.b.}$		$\bar{y}_{...}$

The two-way ANOVA is used to perform the hypotheses test, which will be the following:

1. $H_0 : \tau_1 = \tau_2 = \ldots = \tau_a = 0$ (no main effect of factor A)

   H_1 : at least one $\tau_i \neq 0$

2. $H_0 : \beta_1 = \beta_2 = ... = \beta_b = 0$ (no main effect of factor B)

   H_1 : at least one $\beta_j \neq 0$

3. $H_0 : (\tau\beta)_{11} = (\tau\beta)_{12} = ... = (\tau\beta)_{ab} = 0$ (no interaction)

   H_1 : at least one $(\tau\beta)_{ij} \neq 0$

The ANOVA decomposes the total variability in component parts and the test of hypotheses is based on the comparison between the various elements. The total variability is described by the total sum of squares of the observations, which can be written as

$$SS_T = SS_A + SS_B + SS_{AB} + SS_E \qquad (3.36)$$

The computing formulas for each of the components are

$$SS_T = \sum_{i=1}^{a}\sum_{j=1}^{b}\sum_{k=1}^{n} y_{ijk}^2 - \frac{y_{...}^2}{abn} \qquad (3.37)$$

$$SS_A = \sum_{i=1}^{a} \frac{y_{i..}^2}{bn} - \frac{y_{...}^2}{abn} \qquad (3.38) \qquad SS_B = \sum_{j=1}^{b} \frac{y_{.j.}^2}{an} - \frac{y_{...}^2}{abn} \qquad (3.39)$$

$$SS_{AB} = \sum_{i=1}^{a}\sum_{j=1}^{b} \frac{y_{ij.}^2}{n} - \frac{y_{...}^2}{abn} - SS_A - SS_B \qquad SS_E = SS_T - SS_A - SS_B - SS_{AB} \quad (3.41)$$

$$(3.40)$$

The ratio between the sum of squares and the number of degrees of freedom is the mean square for treatments and it is an estimator for σ^2 if H_0 is true. The error mean square is an estimator for σ^2 whether H_0 is true or not. The calculation expressions for the mean square of each factor, interaction and error are presented in Table 3.3.

To test if the row factors have no main effect, the ratio (3.42) must be used. It has an F-distribution with $a-1$ and $ab(n-1)$ degrees of freedom if H_0 is true. If the condition (3.43) is met, the null hypotheses of no difference in treatment means is accepted with a $100(1-\alpha)\%$ confidence interval, where α defines the sensitivity of the test.

$$F_0 = \frac{MS_A}{MS_E} \qquad (3.42)$$

$$F_0 < f_{\alpha, a-1, ab(n-1)} \tag{3.43}$$

The previous condition, for the row factors, is similar to the column factors and inter-action. A summary is presented in Table 3.3, in the same the ANOVA table results are usually presented.

TABLE 3.3: Two-way ANOVA table

Source of variation	Sum of Squares	Degrees of Freedom	Mean Square	F_0
A treatments	SS_A	$a-1$	$MS_A = \dfrac{SS_A}{a-1}$	$\dfrac{MS_A}{MS_E}$
B treatments	SS_B	$b-1$	$MS_B = \dfrac{SS_B}{b-1}$	$\dfrac{MS_B}{MS_E}$
Interaction	SS_{AB}	$(a-1)(b-1)$	$MS_{AB} = \dfrac{SS_{AB}}{(a-1)(b-1)}$	$\dfrac{MS_{AB}}{MS_E}$
Error	SS_E	$ab(n-1)$	$MS_E = \dfrac{SS_E}{ab(n-1)}$	
Total	SS_T	$abn-1$		

In order to understand the effect the factors have on the algorithm's response, the effects estimate is calculated, according to expressions (3.44), (3.45) and (3.46) where $A_x B_y$ corresponds to the algorithm results when factor A is on the x level and factor B is on y value (x and y being *high* or *low*).

$$A = \frac{1}{2 \cdot n} \left[-\sum_n A_{low} B_{low} + \sum_n A_{high} B_{low} - \sum_n A_{low} B_{high} + \sum_n A_{high} B_{high} \right] \tag{3.44}$$

$$B = \frac{1}{2 \cdot n} \left[-\sum_n A_{low} B_{low} - \sum_n A_{high} B_{low} + \sum_n A_{low} B_{high} + \sum_n A_{high} B_{high} \right] \tag{3.45}$$

$$AB = \frac{1}{2 \cdot n} \left[\sum_n A_{low} B_{low} - \sum_n A_{high} B_{low} - \sum_n A_{low} B_{high} + \sum_n A_{high} B_{high} \right] \tag{3.46}$$

As it can be deducted from the formulas, the signs for each of the sums depends on the level of the factors. For the estimate of the main effects, the signal is negative when the results correspond to the factors low value and positive when the result corresponds to the high value of the factor. In the interaction estimate, the signal is positive if both the results are for the factors on the same level and negative if the values are different.

If the value of the estimate is positive, the response of the algorithm with the factor has a positive direction, meaning that high values of the factor increase the algorithm's response. On the other hand, if the estimate is negative, the direction is negative and high values of the factor result in a lower response from the algorithm.

Knowing the estimate it is possible to adjust the maximum or minimum value of the factors to get a more appropriate result for the algorithm.

Chapter 4

Case Study

4.1 The IEEE 118-bus system

The IEEE 118-bus system represents a part of the USA power system as in December 1962.

The transmission system of the IEEE 118-bus system is composed of 118 buses and 186 lines, with 3 voltage levels, 138 kV, 345 kV and 161 kV. 54 generators are present, making a total installed capacity of 4377.4 MW. The peak load is 4242 MW and 1438 Mvar allocated on 54 buses. There are 9 transformers with taps and its settings were considered in the interval [0.95; 1.05] pu, totaling 10 intervals. The capacitor banks available reactive power is in the range 0-30 Mvar, also divided in 10 intervals. The PV nodes voltage range as well as the PQ nodes maximum voltage deviation is of 8%. Further information regarding the system can be found in [31].

To the generators already present in the base system, wind generation was added in four buses (37, 50, 58 and 96). The same wind scenarios were assumed for all wind generators as well as the maximum power output of 15 MW. The candidate nodes for capacitor banks, TCR and SVC installation are 20. From these, the subset for STATCOM installation is of 10 buses. Figure 4.1 presents the system with the candidate nodes signaled.

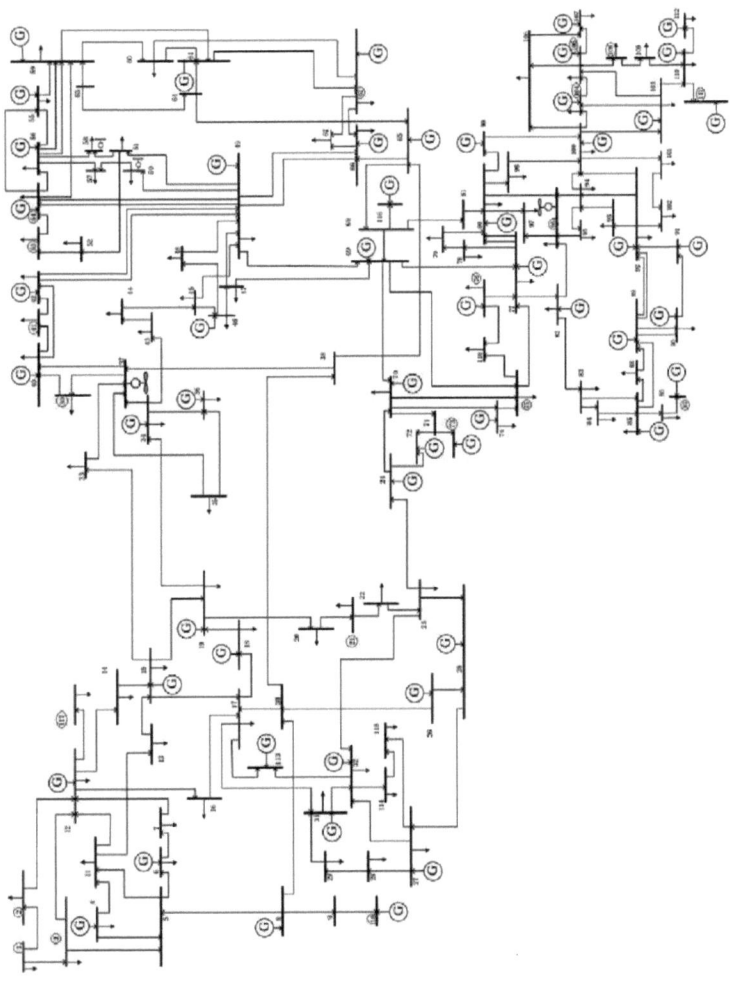

FIGURE 4.1: Modified IEEE 118-bus system.

4.2 Experiments data

The possible locations for the devices installation were defined as the candidate nodes and are highlighted in Figure 4.1. However, STATCOM can only be installed in PQ nodes and therefore the candidate nodes for this kind of devices is a subset of the defined candidate nodes. The candidate nodes for STATCOM are 1, 2, 3, 21, 39, 41, 53, 75, 96, 108 and 117.

The prices of the reactive power compensation devices in all experiments are the ones in Table 4.1. The capacitor banks are cheaper than the FACTS devices but have less compensation benefits, therefore the algorithm must always decide according to the cases between saving in the purchase of the devices or gaining their benefits.

TABLE 4.1: Reactive power compensation devices prices.

Device	Price (€)
Fixed capacitor banks	2700
Switched capacitor banks	4500
TCR	18080
SVC	28910
STATCOM	36138

The ranges for the various devices variables that are not inherent from the network are presented in Table 4.2. The STATCOM voltage limits are the same as the voltage limits on PV nodes.

TABLE 4.2: Devices limits.

Device	Limits
Capacitor banks	10 levels, each of 3Mvar
TCR	0 - 200 Mvar
SVC	-100 - 100 Mvar
Transformer taps	0.95 - 1.05 (levels of 0.01)

The scaling factors of the fitness function limit the algorithm's behaviour. These are presented in Table 4.3.

TABLE 4.3: Scaling factors values.

sfC	0.01
sfE	1
sfV	300
sfM	500
sfQ	300

Finally, the DEEPSO parameters used throughout the experiments are: populations of 160 individuals, a maximum number of generations of 20000 and 500 as the number of unchanged generations for the stopping criteria. The remaining parameters, the communication probability and the mutation rate, will require tuning, which is presented in the next section.

4.3 Parameters tuning

For the parameters tuning, the design of experiment used was the presented in 3.6. The parameters choice was simple, because the ones that most influence DEEPSO (as well as EPSO which is much similar) are widely referred in literature. They are the communication probability and the mutation rate. Therefore, these were the two factors to test and tune. It was also defined that they would range between 0.2 and 0.9 in four steps $[0.2, 0.5, 0.75, 0.9]$. The methodology for parameter tuning is the following: firstly the minimum and maximum values for each factor are 0.2 and 0.9 respectively. 30 runs of the algorithm are performed with the four possible combinations of factors and the results are treated as previously explained. If the F_0 value for any factor does not fulfil the condition 3.43 it means that has main effect in the result and therefore must be changed. The calculated estimate of the effect is analysed and a decision on whether increase the low value or decrease the high is made in order to decrease the algorithm's response, since the aim is minimizing the fitness function. This procedure is repeated until no differences are present.

For both models the F-distribution value used to test the hypotheses is always the same, since the number of degrees of freedom are the same as well as the confidence interval. The value for comparison is:

$$f_{\alpha,a-1,ab(n-1)} = f_{0.05,1,116} = 3.9229 \qquad (4.1)$$

For each run of each model, two tables are presented: one with the values of the factors for the test and other with the ANOVA's F_0 and effects. In both tables, the relevant values are highlighted, the changed from previous tests in the former and the values that do not meet condition 3.43 in the latter.

4.3.1 Deterministic model

TABLE 4.4: Values for test 1 of the deterministic model

(a) Factor values

	Communication probability	Mutation rate
Minimum	0.2	0.2
Maximum	0.9	0.9

(b) ANOVA table.

Source	F_0	Estimate
Columns	*144.984*	36871.7
Rows	0.030	532.7
Interaction	0.210	1402.2

TABLE 4.5: Values for test 2 of the deterministic model

(a) Factor values

	Communication probability	Mutation rate
Minimum	0.2	0.2
Maximum	**0.75**	0.9

(b) ANOVA table

Source	F_0	Estimate
Columns	*197.331*	34588.0
Rows	*6.006*	6034.4
Interaction	*7.862*	6903.9

TABLE 4.6: Values for test 3 of the deterministic model

(a) Factors values

	Communication probability	Mutation rate
Minimum	0.2	0.2
Maximum	**0.5**	**0.75**

(b) ANOVA table.

Source	F_0	Estimate
Columns	*116.618*	22594.6
Rows	0.616	-1642.0
Interaction	0.064	531.4

At this point, the F_0 value for the communication factor is still high. As the estimate is positive, the high value of this factor should decrease further for the next test. However, it is already in its lowest step. Thus, the test will not advance further and the value adopted for the communication factor will be the low value in this last test, 0.2. For the mutation rate, the medium value in the interval will be used: 0.475.

4.3.2 Probabilistic method

TABLE 4.7: Values for test 1 of the probabilistic model

(a) Factors values			(b) ANOVA table.		
	Communication probability	Mutation rate	Source	F_0	Estimate
			Columns	*5.325*	2296.6
Minimum	0.2	0.2	Rows	1.331	1148.0
Maximum	0.9	0.9	Interaction	3.229	19807.5

TABLE 4.8: Values for test 2 of the probabilistic model

(a) Factors values.			(b) ANOVA table.		
	Communication probability	Mutation rate	Source	F_0	Estimate
			Columns	*13.015*	513.1
Minimum	0.2	0.2	Rows	0.089	42.6
Maximum	**0.75**	0.9	Interaction	*17.511*	595.1

TABLE 4.9: Values for test 3 of the probabilistic model

(a) Factors values.			(b) ANOVA table.		
	Communication probability	Mutation rate	Source	F_0	Estimate
			Columns	*4.696*	254.3
Minimum	0.2	0.2	Rows	*13.892*	-437.4
Maximum	**0.5**	0.9	Interaction	2.611	189.6

The communication factor high value cannot be lowered any more, therefore only the mutation rate is altered.

TABLE 4.10: Values for test 4 of the probabilistic model

(a) Factors values.			(b) ANOVA table.		
	Communication probability	Mutation rate	Source	F_0	Estimate
			Columns	1.022	129.5
Minimum	0.2	**0.5**	Rows	1.997	-181.0
Maximum	0.5	0.9	Interaction	2.957	220.3

Since after test 4 all effects meet the necessary condition, the values that will be used in the algorithm will be the middle values of the intervals, 0.35 for the communication probability and 0.7 for the mutation rate.

4.4 Results

4.4.1 Deterministic model results

The main objective of the deterministic model was to determine if RPP based on extreme deterministic cases is practicable. In order to accomplish this, four runs of the algorithm were performed, each with a single scenario, with the combinations of load and wind generation presented in Table 4.11.

TABLE 4.11: Deterministic scenarios description.

		Wind generation	
		low	high
Load	off-peak	S1	S2
	peak	S3	S4

To verify the compatibility between the solutions for the different scenarios, the locations of newly installed devices in each scenario must be compared. These results are organized and presented in Table 4.12. In this experiment, no STATCOM was installed which is why it is not presented. This fact is due to the high cost of this device, that isn't justifiable with only one scenario.

The cases highlighted in the table are the ones that make the use of the deterministic cases impracticable. In the case of capacitor banks, if only cases like bus 41 and 53 were present there would be no problem because the installation of fixed and switched capacitor banks would solve them. However, bus 39 has installed devices in some scenarios and not installed in others, which demonstrates a incompatibility between the scenarios solutions. The same happens in the highlighted nodes with installation of TCR and SVC devices, which only happens in some scenarios.

The discrepancies between the results for the different scenarios demonstrate that RPP based on extreme cases of load and wind generation is not a very good approach. If one was only to use a single scenario, it would not be adequate to other extreme scenario. On the other hand, if one were to use more than one scenario, they would not be compatible between each other.

This incompatibility between different deterministic scenarios was the main motivation for further development of the method in order to combine different scenarios, each with

TABLE 4.12: Installed devices in the deterministic model.

candidate nodes		1	2	3	10	21	39	41	53	54	62	73	75	76	87	96	104	105	108	111	117
Capacitor banks	S1	9	2	4	0	4	4	3	5	0	0	0	7	0	0	7	0	0	0	0	2
	S2	9	2	4	0	3	5	3	4	0	0	0	6	0	0	6	0	0	0	0	2
	S3	9	3	8	0	6	7	3	6	0	0	0	8	0	0	9	0	0	0	0	3
	S4	4	3	7	0	5	0	3	1	0	0	0	7	0	0	4	0	0	0	0	3
TCR	S1	0	0	0	0	0	0	0	0	0	0	0	0	0	0	0	0	0	0	0	0
	S2	0	0	0	0	0	0	0	1	0	0	0	0	0	0	1	0	0	0	0	0
	S3	0	0	0	0	0	0	0	0	0	0	0	0	0	0	0	0	0	0	0	0
	S4	1	0	0	0	0	0	0	1	0	0	0	0	0	0	1	0	0	0	0	0
SVC	S1	0	0	0	0	0	0	0	0	0	0	0	0	0	0	0	0	0	0	0	0
	S2	0	0	0	0	0	0	0	0	0	0	0	0	0	0	0	0	0	0	0	0
	S3	0	0	0	0	0	0	0	0	0	0	0	0	0	0	0	0	0	0	0	0
	S4	0	0	0	0	0	1	0	0	0	0	0	0	0	0	0	0	0	0	0	0

a probability associated. This is the probabilistic method and its results are presented next.

4.4.2 Probabilistic model results

The experiments performed with the probabilistic method have two main objectives: demonstrate its improvement over the deterministic method and show how the RPP problem objectives are contrary and provide a trade-off analysis of the results.

The scenarios used in the experiments for this model are the ones presented in 3.3 and Table 4.13 shows the denomination of each scenario and their probability.

TABLE 4.13: Probabilisitc scenarios denomination and probability.

		Wind generation		
		low	medium	high
Load	off-peak	S1: 8.97%	S2: 14.79%	S3: 1.24%
	normal	S4: 17.94%	S5: 29.58%	S6: 2.48%
	peak	S7: 8.97%	S8: 14.79%	S9: 1.24%

The model was tested with a maximum voltage deviation of 6%, a value close to the maximum allowed in the network. The solution obtained consisted in the installation of 7 fixed and 13 switched capacitor banks (270 Mvar of fixed capacity and 231 Mvar of additional switched capacity), 1 TCR and 1 STATCOM. Table 4.14 presents the capacitor banks levels on all scenarios and nodes and the nodes with switched capacitor banks have a different background colour. The TCR is installed in bus 87 and the STATCOM on bus 117, which are buses with the capacitor banks at the maximum level for all the scenarios and thus the FACTS are used as reinforcement.

TABLE 4.14: Probabilistic model capacitor banks placement for VDP of 6%.

	1	2	3	10	21	39	41	53	54	62	73	75	76	87	96	104	105	108	111	117
S1	9	3	3	2	4	4	3	5	7	9	10	7	8	10	7	8	10	1	10	10
S2	9	3	3	10	3	4	3	7	8	10	10	7	10	10	10	10	4	1	10	10
S3	9	3	3	1	3	5	3	10	2	10	10	7	3	10	10	1	3	1	10	10
S4	8	3	5	10	5	5	3	5	10	10	10	9	10	10	8	10	10	1	10	10
S5	8	3	5	10	4	6	3	7	9	3	10	9	10	10	10	6	10	1	10	10
S6	9	3	4	8	4	6	3	10	10	9	10	9	9	10	10	10	10	1	10	10
S7	9	3	7	10	6	7	3	6	10	10	10	8	10	10	9	2	6	1	10	10
S8	9	3	7	4	6	7	3	8	10	0	10	9	10	10	10	2	10	1	10	10
S9	10	3	7	1	1	7	3	10	10	8	10	10	1	10	10	10	10	1	10	10

The solution reached results in the fitness values presented in Table 4.15. The system power losses decrease from 125.575 MW to 124.062 MW, corresponding to annual energy savings costs of 253 018.10€. From the fitness function values, it is also possible to conclude that neither of the constraints are violated.

TABLE 4.15: Fitness function values for probabilistic model with VDP 6%.

Investment cost	131 618,00
Losses cost	20 928 047,06
Voltage deviation cost	0,00
Reactive generated power deviation cost	0,00
Fitness function value	**20 929 363,24**

The voltage profile on the PQ nodes is presented in Figure 4.2, including the maximum and minimum voltages allowed. These are the voltages more interesting to analyse since they are the ones controlled by penalties. All voltages are included in the ±6% limit over the nominal voltage.

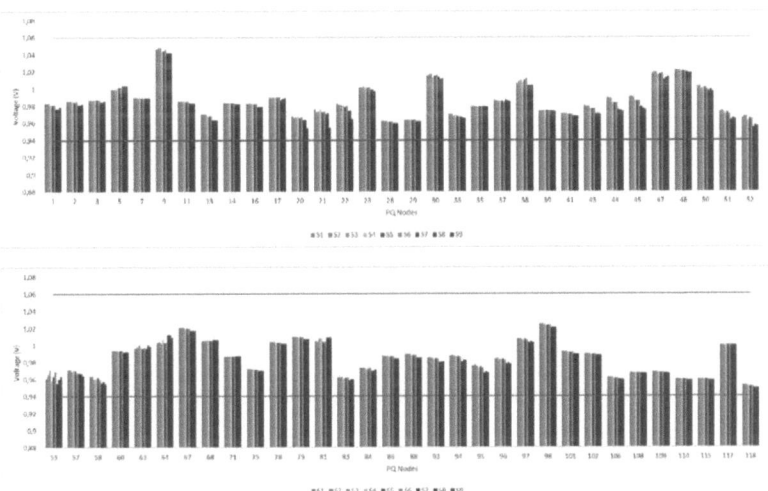

FIGURE 4.2: Voltage profile of PQ nodes for maximum voltage deviation of 6%.

Figure 4.3 presents the algorithm convergence rate in this case. It can be observed that the algorithm reaches rapidly the neighbourhood of the solution and then there is a finer tuning until it finally meets the stopping criteria.

FIGURE 4.3: Probabilistic method algorithm convergence, with VDP of 6%.

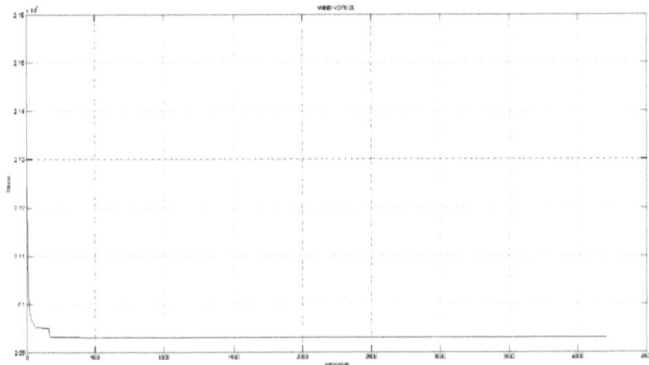

The model was also tested with a maximum voltage deviation of 3% to understand its behaviour with such a difficult case. The solution required the installation of 20 switched capacitor banks, with a fixed capacity of 174 Mvar and 345 Mvar of additional switched capacity, 2 TCR (on buses 73 and 111), 3 SVC (on buses 10, 54 and 76) and 1 STATCOM (on bus 117).

Due to the hard constraint present in this case, the algorithm is forced to install more expensive devices, in order to be able to minimize the voltage deviation as much as possible. This is also demonstrated by the fitness values from the solution, which have higher values for the losses and investment cost. The main factors that influence this behaviour are the scaling factors that were defined and give preference to reduce voltage deviations at the expenses of higher losses and investment costs.

The active power losses in this case decrease to 124.11 MW, the annual savings cost being 245 827,55€.

TABLE 4.16: Results for probabilistic model with VDP 6%.

Investment cost	249 028,00
Losses cost	20 935 237,61
Voltage deviation cost	476,89
Reactive generated power deviation cost	0,00
Fitness function value	**21 080 794,47**

Although the algorithm's efforts to prevent voltage deviations higher than 3%, we may conclude that this was not possible, both from the values of the fitness function voltage deviation cost and the voltage profile on the PQ nodes (Figure 4.4). In reality, the maximum voltage deviation in this case is a very low value that is seldom practised in real systems and was used to understand the conduct of the algorithm.

Comparing the voltage profiles for both maximum voltage deviation of 6% and 3% it is possible to conclude that they are very similar and therefore that is an optimal voltage profile for the system.

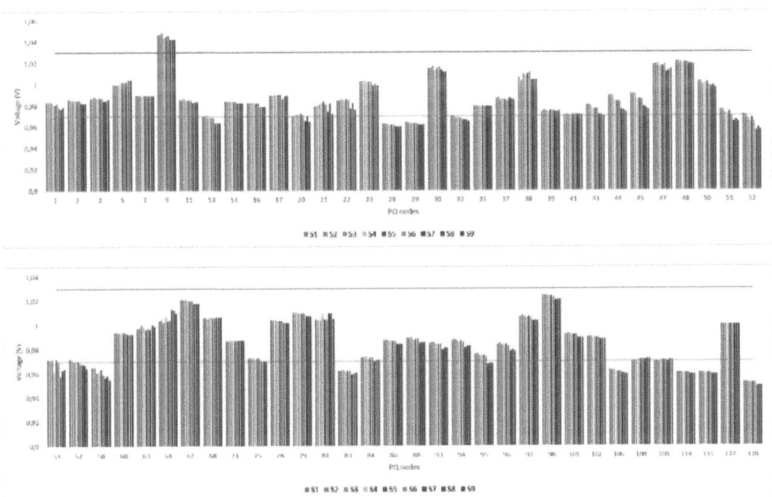

FIGURE 4.4: Voltage profile of PQ nodes for maximum voltage deviation of 3%.

The algorithm was run with a third maximum voltage deviation allowed, with value 8%. These results were combined with the previous ones to study how the algorithm handles the conflicting objectives of the fitness function. Figure 4.5 presents the conflict between the maximum allowed voltage deviation and the active power losses. Both objectives should be minimized but, as it is clear by the graph, there is no optimum solution.

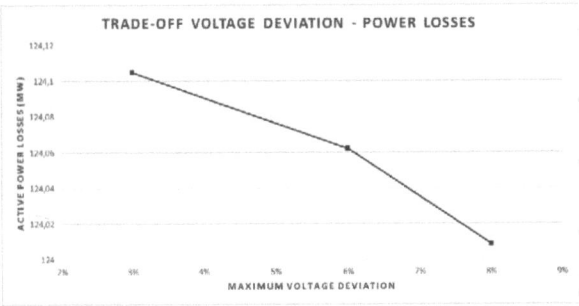

FIGURE 4.5: Maximum voltage allowed vs. active power losses

Chapter 5

Conclusions

The major achievement accomplished on the course of this research work is the development of a probabilistic method to solve the RPP problem on power systems with uncertain load and wind generation with the use of the DEEPSO, a recent and powerful algorithm, to perform the computations.

From the perspective of the RPP problem, and power systems planning in general, this work applies the developed method to deterministic scenarios and successfully demonstrates that planning the purchase and installation of new devices in the network based solely on these is not accurate nor reliable.

From the standpoint of the proposed method, a noteworthy contribution is given on the development of probabilistic model to solve the RPP problem. The present work presents a probabilistic method to optimize the reactive power in systems with uncertain load and wind generation. Therefore the solutions are for more realistic systems with load and wind generation scenarios that have associated probabilities.

The kind of devices used in the RPP problem was also expanded in this research work, by combining hybrid capacitor banks with three different FACTS devices, each with different benefits and costs.

An understanding of the trade-off present between the opposing objectives of the problem is also provided by this thesis

The developed method and corresponding results have room for further evolution, by the inclusion of contingency scenarios. This would allow to broaden the planning solutions even further and prepare the system for a better response to emergency situations.

Also, the results depend on the parameter costs and, with different values, the results could be different. This work may also be a starting point for a study of the sensitivity of the solutions to cost variations.

Bibliography

[1] The wind power - wind turbines and wind farms database, 2005-2014.

[2] X. Wu. Reactive Power Compensation Based on FACTs Devices.

[3] J. Paulo T. Saraiva, J. L. P. Pereira da Silva, and M. T. P. de Leão. *Mercados de Electricidade - Regulação e Tarifação de Uso das Redes.* 1 edition, 2002.

[4] T. Ackermann and L. Soder. An overview of wind energy status-2002. *Renewable and Sustainable Energy Reviews*, 6(1-2):67–128, 2002.

[5] N. G. Hingorani and L. Gyugyi. *Understanding FACTS: Concepts and Technology of Flexible AC Transmission Systems.* Wiley, 2000.

[6] W. Zhang, F. Li, and L. Tolbert. Review of reactive power planning: objectives, constraints, and algorithms. In *Transmission and Distribution Conference and Exposition, 2008. IEEE/PES*, pages 1–1, April 2008.

[7] K. Iba, H. Suzuki, K.-I. Suzuki, and K. Suzuki. Practical reactive power allocation/operation planning using successive linear programming. *Power Systems, IEEE Transactions on*, 3(2):558–566, May 1988.

[8] B. Kermanshahi, K. Takahashi, and Yicheng Zhou. Optimal operation and allocation of reactive power resource considering static voltage stability. In *Power System Technology, 1998. Proceedings. POWERCON '98. 1998 International Conference on*, volume 2, pages 1473–1477 vol.2, Aug 1998.

[9] N.P. Padhy, M.A. Abdel-Moamen, and B.J. Praveen Kumar. Optimal location and initial parameter settings of multiple tcscs for reactive power planning using genetic algorithms. In *Power Engineering Society General Meeting, 2004. IEEE*, pages 1110–1114 Vol.1, June 2004.

[10] V. Miranda, N. W. Oo, and J. N. Fidalgo. Experimenting in the Optimal Capacitor Placement and Control Problem with Hybrid Mathematical-Genetic Algorithms. In *ISAP -INTERNATIONAL CONFERENCE; Intelligent system application to power systems International conference, Intelligent system application to power systems,* number 2, pages 188–196, 2001.

[11] L. L. Lai and J. T. Ma. Application of evolutionary programming to reactive power planning-comparison with nonlinear programming approach. *Power Systems, IEEE Transactions on,* 12(1):198–206, Feb 1997.

[12] V. Gopalakrishnan, P. Thirunavukkarasu, and R. Prasanna. Reactive power planning using hybrid evolutionary programming method. In *Power Systems Conference and Exposition, 2004. IEEE PES,* pages 1319–1323 vol.3, Oct 2004.

[13] C. Chen, H. Lee, and W. Tsai. Optimal reactive power planning using genetic algorithm. In *Systems, Man and Cybernetics, 2006. SMC '06. IEEE International Conference on,* volume 6, pages 5275–5279, Oct 2006.

[14] H. Keko, A.J. Duque, and V. Miranda. A multiple scenario security constrained reactive power planning tool using epso. In *Intelligent Systems Applications to Power Systems, 2007. ISAP 2007. International Conference on,* pages 1–6, Nov 2007.

[15] M. Eghbal, E. E. El-Araby, N. Yorino, and Y. Zoka. Application of metaheuristic methods to reactive power planning: a comparative study for ga, pso and epso. In *Systems, Man and Cybernetics, 2007. ISIC. IEEE International Conference on,* pages 3755–3760, Oct 2007.

[16] Y. L Chen and Y.-L. Ke. Multi-objective var planning for large-scale power systems using projection-based two-layer simulated annealing algorithms. *Generation, Transmission and Distribution, IEE Proceedings-,* 151(4):555–560, July 2004.

[17] P. Rajkumar and D. Devaraj. Adaptive particle swarm optimization approach for optimal reactive power planning. In *Power System Technology and IEEE Power India Conference, 2008. POWERCON 2008. Joint International Conference on,* pages 1–7, Oct 2008.

[18] Y. Mao and M. li. Optimal reactive power planning based on simulated annealing particle swarm algorithm considering static voltage stability. 1:106–110, Oct 2008.

[19] Z. Yiqin. Optimal reactive power planning based on improved tabu search algorithm. In *Electrical and Control Engineering (ICECE), 2010 International Conference on*, pages 3945–3948, June 2010.

[20] Y. Wang, F. Li, Q. Wan, and H. Chen. Multi-objective reactive power planning based on fuzzy clustering and learning automata. In *Power System Technology (POWERCON), 2010 International Conference on*, pages 1–7, Oct 2010.

[21] Y. Hong and K. Pen. Optimal var planning considering intermittent wind power using markov model and quantum evolutionary algorithm. *Power Delivery, IEEE Transactions on*, 25(4):2987–2996, Oct 2010.

[22] A. a. Cuello-Reyna and J. R. Cedeno-Maldonado. Combined Analytic Hierarchical Process-Differential Evolution Approach for Optimal Reactive Power Planning. *2006 International Conference on Probabilistic Methods Applied to Power Systems*, pages 1–8, June 2006.

[23] M. Rahmani, M. Rashidinejad, E.M. Carreno, and R.A. Romero. A combinatorial approach for transmission expansion amp; reactive power planning. In *Transmission and Distribution Conference and Exposition: Latin America (T D-LA), 2010 IEEE/PES*, pages 529–536, Nov 2010.

[24] S. Kanokbannakorn and K. Audomvongseree. Cost-based reactive power planning in distribution system considering reliability. In *Electrical Engineering/Electronics, Computer, Telecommunications and Information Technology (ECTI-CON), 2011 8th International Conference on*, pages 824–827, May 2011.

[25] M. Alonso, H. Amaris, and C. Alvarez-Ortega. A multiobjective approach for reactive power planning in networks with wind power generation. *Renewable Energy*, 37(1):180–191, January 2012.

[26] M. Niu and Z. Xu. Reactive power planning for transmission grids with wind power penetration. pages 1–5, May 2012.

[27] V. Miranda and N. Fonseca. Epso - best-of-two-worlds meta-heuristic applied to power system problems. In *Evolutionary Computation, 2002. CEC '02. Proceedings of the 2002 Congress on*, volume 2, pages 1080–1085, 2002.

[28] V. Miranda, H. Keko, and Á.J. Duque. Stochastic star communication topology in evolutionary particle swarms (epso). *Intern. Journal of Computational Intelligence Research*, 4:105–116, 2008.

[29] V. Miranda and R. Alves. Differential Evolutionary Particle Swarm Optimization (DEEPSO): a successful hybrid. 2013.

[30] D.C. Montgomery and G.C. Runger. *Applied statistics and probability for engineers.* Wiley, 1999.

[31] R. Christie. Power systems test case archive, May 1993.

Printed by Books on Demand GmbH, Norderstedt / Germany